# UAS
## Pilot Log

## Expanded Edition

Unmanned Aircraft Systems
Logbook for Drone Pilots & Operators

created by
**droneprep**
droneprep.com
@droneprep

By publishing this book, neither the authors nor the publisher are engaged in rendering legal or other professional services. If any such assistance is required, the services of a qualified professional or should be sought. The authors and publisher will not be responsible for any liability, loss, or risk incurred as a result of the use and application of any of the information contained in this book.

# Logbook Onwer Information

**First and Last Name:** _____

**Contact Information:**

    E-mail Address: _____

    Physical Address: _____

                              _____

                              _____

                              _____

    Phone: _____

**Certificates & Ratings:** _____

                              _____

                              _____

                              _____

**Aircrafts Owned:** _____

                              _____

                              _____

# How To Use This Log Book

Congratulations on your decision to purchase Unmanned Aircraft Systems: Drone Pilot and Operator Log Expanded Edition. After receiving extensive user feedback and testing, we are excited to build on the success of the standard droneprep pilot log with expanded features to assist the routine operations for pilots and operators of unmanned aircraft systems and drones. This log book captures all the elements contained in droneprep's standard log book but offers additional and expanded tools to enhance the flying and record keeping experience.

Using this logbook, pilots will be able to distill complex procedures and note taking with simple, easy-to-understand entry pages that can be maintained by any drone operator, regardless of skill level or experience. Added spacing, graphs, notational areas, and commentary fields make this Expanded Edition a flexible and powerful record that will serve as both a tool to enhance the flight experience and a superb record of exactly what happened on the day of each flight.

To get started, please follow this brief tutorial that explains how to use each section of the log book so you can maximize your flying experience.

## 1. Flight Session Information

One of the key elements of Unmanned Aircraft Systems: Drone Pilot and Operator Log Expanded Edition is the use of two-pages for each flight session. You will notice that both the left and right-hand side of the page should be used to capture the details related to a single flight session.

The Flight Session Information section provides an opportunity to describe each flying session. Since a given flying session may involve multiple takeoffs and landings, we describe below how to reflect the flying time for each interval below under the "Flight Notes") section. We therefore recommend you identify the "Flight Session" with a particular identification or number so it is unique.

I. FLIGHT ID / NO.: _____       Month _____ Day _____ Year _____

FLIGHT LOCATION: _____       Weather: _____

For example, under **I. Flight ID / NO.**, we suggest establishing your own numbering system so you can differentiate one flying session from the next. This will help you maintain a process driven approach to record keeping that is both standardized and applicable to future flights.

Under **Flight Location** you can be as detailed as you like. The goal is to simply memorialize the starting point for your flight along with the **Date** and **Weather** conditions. We recommend populating every field as these details become useful later as you start to analyze data across multiple flights.

## 2. Aircraft and Crew Data

The Aircraft and Crew Data section is straightforward and designed to capture details about the **Unmanned Aircraft System** used in a flight session's operations. We suggest this to be as detailed as necessary to capture the identity of craft so there is a reasonable level of uniqueness recorded, including the craft's **manufacturer** and **model number**.

UNMANNED AIRCRAFT SYSTEM:                          CREW:

MANUFACTURER: _____        PILOT: _____

MODEL NUMBER: _____        SPOTTER: _____

This section also captures personnel who are involved in a given flight session. The requirement to detail the identity of both the **Pilot in Command** and **Spotter** highlights the importance of multi-person approaches to flying. The use of a spotter during flight operations lends itself to greater safety and more successful operations.

## 3. Preflight Checklist

Every Pilot in Command should make preflight checks and procedures part of the regular routine before each flight. While every pilot may approach preflight checks differently, we provide an opportunity for you to confirm that some of the most common aircraft checks are completed. Here is a sample of the section:

**PREFLIGHT CHECKLIST:**

| | | |
|---|---|---|
| ☐ Batteries Charged & Secure | ☐ Props OK & Tight | ☐ Compass Calibration |
| ☐ Aircraft Hardware OK | ☐ Software / Firmware Update | ☐ Camera / FPV ON |
| ☐ Equipment & Gear OK | ☐ Transmitter Control Power ON | ☐ Satellite Connection |
| ☐ Transmitter Controls OK | ☐ Aircraft Power ON | ☐ Applications / Other Systems ON |

**OPERATIONAL CONDITIONS & PREFLIGHT NOTES:**

As shown, you also have an area to detail any **Operational Conditions and Preflight Notes** to memorialize any activities or circumstances that may relevant to the day's flight or safety operations.

Please use the above checks as a guide and not necessarily an exhaustive list. Remember, the goal is to make sure you, your crew, your craft and the people around you are safe at all times. Preflight are one way you can achieve this, but ultimately it's up to you to make sure

## 4. Session Flight Intervals

As a UAS or drone operator, the reality is that you may take your craft up multiple times in a single day's session. This may be due to a handful of reasons, including battery life, video recording space or those unfortunate premature and unplanned landings. Unlike other log books, which are based on traditional airplane style operations, we designed this log to reflect the practical realities of the UAS and drone pilot who operate small crafts.

| SESSION FLIGHT INTERVALS | FLIGHT TIMES | | | REMARKS |
|---|---|---|---|---|
| | START | STOP | TOTAL | PROCEDURES & MANEUVERS |
| 1. | | | | |
| 2. | | | | |
| 3. | | | | |
| 4. | | | | |
| 5. | | | | |
| 6. | | | | |
| 7. | | | | |
| 8. | | | | |
| 9. | | | | |
| 10. | | | | |
| TOTAL HOURS FOR SESSION | | | | |
| TOTAL FORWARD | | | | |
| TOTAL TO DATE | | | | |

To populate the **Session Flight Intervals** table, start by entering details on line number 1 for your first flight within a given session. Adjacent to the number 1 (or numbers 2 through 10), you may enter a name, number or any other unique identifier to represent that specific interval within the session.

For **START**, enter either the time of day or, if using a stopwatch method of timing, the starting time. For **STOP**, enter the corresponding time of day or stopping time point. The difference between STOP and START is then reflected in **TOTAL** for that particular flight interval. Repeat the same log entry practice for up to seven intervals. Should you need more intervals for the same flight session, simply use another page. Once your intervals and your flight session is completed for the day, sum the TOTAL lines together and enter the value in the **TOTAL HOURS FOR SESSION** box.

Underneath the TOTAL HOURS FOR SESSION box is **TOTAL FORWARD**, which is the **TOTAL TO DATE** value from the immediately prior flight session. If you are making your first very first flight session entry, there is no TOTAL FORWARD. **TOTAL TO DATE** is therefore the sum of TOTAL HOURS FOR SESSION plus the TOTAL FORWARD.

Also included are the **REMARKS, PROCEDURES & MANEUVERS** column where you can notate specific aspects of each flight session.

## 5. Flight Map

The **Flight Map** affords the operator a chance to visually depict or draw various elements from the flight session. This section is optional and should be flexible to the needs of the operator, but we have found it can add value from a preflight planning or post flight analysis standpoint. For example, you may want to plot your flight path and any relevant way points. Alternatively, you may also want to use this area to help you conceptualize the flying field and any possible obstructions or elevation changes. Ultimately, you may use this area as you see fit to promote a safe and effective flight and to capture a record of exactly what happened during your flight session.

**FLIGHT MAP**

# 6. Other Entries

The remaining entries simply allow you to enter any **Post flight Notes** and **Journal Entries**. After your flight, take a few minutes to reflect on what happened and whether any of the session's events are worth recording for future reference. Once you have completed your entries, please sign the bottom of the page—along with the Spotter—to certify that your entries are true and correct.

# Log Entry Pages

# I. FLIGHT ID / NO.: _____

FLIGHT LOCATION: _____     Weather: _____

UNMANNED AIRCRAFT SYSTEM:                    CREW:

MANUFACTURER: _____        PILOT: _____

MODEL NUMBER: _____        SPOTTER: _____

PREFLIGHT CHECKLIST:

☐ Batteries Charged & Secure      ☐ Props OK & Tight            ☐ Compass Calibration
☐ Aircraft Hardware OK            ☐ Software / Firmware Update   ☐ Camera / FPV ON
☐ Equipment & Gear OK             ☐ Transmitter Control Power ON ☐ Satellite Connection
☐ Transmitter Controls OK         ☐ Aircraft Power ON           ☐ Applications / Other Systems ON

OPERATIONAL CONDITIONS & PREFLIGHT NOTES:

| SESSION FLIGHT INTERVALS | FLIGHT TIMES | | | REMARKS |
| --- | --- | --- | --- | --- |
| | START | STOP | TOTAL | PROCEDURES & MANEUVERS |
| 1. | | | | |
| 2. | | | | |
| 3. | | | | |
| 4. | | | | |
| 5. | | | | |
| 6. | | | | |
| 7. | | | | |
| 8. | | | | |
| 9. | | | | |
| 10. | | | | |
| TOTAL HOURS FOR SESSION | | | | |
| TOTAL FORWARD | | | | |
| TOTAL TO DATE | | | | |

# II. FLIGHT MAP

Month _____ Day _____ Year _____

POSTFLIGHT NOTES / JOURNAL ENTRIES:

I CERTIFY THAT THE FOREGOING ENTRIES ARE TRUE AND CORRECT:

PILOT: _____ SPOTTER: _____

# I. FLIGHT ID / NO.: _____

FLIGHT LOCATION: _____     Weather: _____

UNMANNED AIRCRAFT SYSTEM:                    CREW:

MANUFACTURER: _____          PILOT: _____

MODEL NUMBER: _____          SPOTTER: _____

## PREFLIGHT CHECKLIST:

☐ Batteries Charged & Secure      ☐ Props OK & Tight              ☐ Compass Calibration
☐ Aircraft Hardware OK            ☐ Software / Firmware Update     ☐ Camera / FPV ON
☐ Equipment & Gear OK             ☐ Transmitter Control Power ON   ☐ Satellite Connection
☐ Transmitter Controls OK         ☐ Aircraft Power ON              ☐ Applications / Other Systems ON

## OPERATIONAL CONDITIONS & PREFLIGHT NOTES:

| SESSION FLIGHT INTERVALS | FLIGHT TIMES | | | REMARKS |
|---|---|---|---|---|
|  | START | STOP | TOTAL | PROCEDURES & MANEUVERS |
| 1. |  |  |  |  |
| 2. |  |  |  |  |
| 3. |  |  |  |  |
| 4. |  |  |  |  |
| 5. |  |  |  |  |
| 6. |  |  |  |  |
| 7. |  |  |  |  |
| 8. |  |  |  |  |
| 9. |  |  |  |  |
| 10. |  |  |  |  |
| TOTAL HOURS FOR SESSION |  |  |  |  |
| TOTAL FORWARD |  |  |  |  |
| TOTAL TO DATE |  |  |  |  |

## II. FLIGHT MAP

Month _____ Day _____ Year _____

**POSTFLIGHT NOTES / JOURNAL ENTRIES:**

I CERTIFY THAT THE FOREGOING ENTRIES ARE TRUE AND CORRECT:

PILOT: _____ SPOTTER: _____

# I. FLIGHT ID / NO.: _____

FLIGHT LOCATION: _____     Weather: _____

UNMANNED AIRCRAFT SYSTEM:                   CREW:

MANUFACTURER: _____           PILOT: _____

MODEL NUMBER: _____           SPOTTER: _____

PREFLIGHT CHECKLIST:

☐  Batteries Charged & Secure     ☐  Props OK & Tight                ☐  Compass Calibration
☐  Aircraft Hardware OK           ☐  Software / Firmware Update      ☐  Camera / FPV ON
☐  Equipment & Gear OK            ☐  Transmitter Control Power ON    ☐  Satellite Connection
☐  Transmitter Controls OK        ☐  Aircraft Power ON               ☐  Applications / Other Systems ON

OPERATIONAL CONDITIONS & PREFLIGHT NOTES:

| SESSION FLIGHT INTERVALS | FLIGHT TIMES | | | REMARKS |
| --- | --- | --- | --- | --- |
| | START | STOP | TOTAL | PROCEDURES & MANEUVERS |
| 1. | | | | |
| 2. | | | | |
| 3. | | | | |
| 4. | | | | |
| 5. | | | | |
| 6. | | | | |
| 7. | | | | |
| 8. | | | | |
| 9. | | | | |
| 10. | | | | |
| TOTAL HOURS FOR SESSION | | | | |
| TOTAL FORWARD | | | | |
| TOTAL TO DATE | | | | |

## II. FLIGHT MAP

Month _____ Day _____ Year _____

**POSTFLIGHT NOTES / JOURNAL ENTRIES:**

I CERTIFY THAT THE FOREGOING ENTRIES ARE TRUE AND CORRECT:

PILOT: _____     SPOTTER: _____

# I. FLIGHT ID / NO.: _____

FLIGHT LOCATION: _____    Weather: _____

UNMANNED AIRCRAFT SYSTEM:                 CREW:

MANUFACTURER: _____       PILOT: _____

MODEL NUMBER: _____       SPOTTER: _____

PREFLIGHT CHECKLIST:

☐ Batteries Charged & Secure    ☐ Props OK & Tight              ☐ Compass Calibration
☐ Aircraft Hardware OK          ☐ Software / Firmware Update     ☐ Camera / FPV ON
☐ Equipment & Gear OK           ☐ Transmitter Control Power ON   ☐ Satellite Connection
☐ Transmitter Controls OK       ☐ Aircraft Power ON             ☐ Applications / Other Systems ON

OPERATIONAL CONDITIONS & PREFLIGHT NOTES:

| SESSION FLIGHT INTERVALS | FLIGHT TIMES | | | REMARKS |
| --- | --- | --- | --- | --- |
| | START | STOP | TOTAL | PROCEDURES & MANEUVERS |
| 1. | | | | |
| 2. | | | | |
| 3. | | | | |
| 4. | | | | |
| 5. | | | | |
| 6. | | | | |
| 7. | | | | |
| 8. | | | | |
| 9. | | | | |
| 10. | | | | |
| TOTAL HOURS FOR SESSION | | | | |
| TOTAL FORWARD | | | | |
| TOTAL TO DATE | | | | |

# II. FLIGHT MAP

Month _____ Day _____ Year _____

**POSTFLIGHT NOTES / JOURNAL ENTRIES:**

I CERTIFY THAT THE FOREGOING ENTRIES ARE TRUE AND CORRECT:

PILOT: _____ SPOTTER: _____

# I. FLIGHT ID / NO.: _____

FLIGHT LOCATION: _____        Weather: _____

UNMANNED AIRCRAFT SYSTEM:                        CREW:

MANUFACTURER: _____               PILOT: _____

MODEL NUMBER: _____               SPOTTER: _____

## PREFLIGHT CHECKLIST:

☐  Batteries Charged & Secure      ☐  Props OK & Tight              ☐  Compass Calibration
☐  Aircraft Hardware OK            ☐  Software / Firmware Update     ☐  Camera / FPV ON
☐  Equipment & Gear OK             ☐  Transmitter Control Power ON   ☐  Satellite Connection
☐  Transmitter Controls OK         ☐  Aircraft Power ON             ☐  Applications / Other Systems ON

## OPERATIONAL CONDITIONS & PREFLIGHT NOTES:

| SESSION FLIGHT INTERVALS | FLIGHT TIMES | | | REMARKS |
|---|---|---|---|---|
| | START | STOP | TOTAL | PROCEDURES & MANEUVERS |
| 1. | | | | |
| 2. | | | | |
| 3. | | | | |
| 4. | | | | |
| 5. | | | | |
| 6. | | | | |
| 7. | | | | |
| 8. | | | | |
| 9. | | | | |
| 10. | | | | |
| TOTAL HOURS FOR SESSION | | | | |
| TOTAL FORWARD | | | | |
| TOTAL TO DATE | | | | |

## II. FLIGHT MAP

Month _____ Day _____ Year _____

**POSTFLIGHT NOTES / JOURNAL ENTRIES:**

I CERTIFY THAT THE FOREGOING ENTRIES ARE TRUE AND CORRECT:

PILOT: _____ SPOTTER: _____

# I. FLIGHT ID / NO.: _____

FLIGHT LOCATION: _____     Weather: _____

UNMANNED AIRCRAFT SYSTEM:                      CREW:

MANUFACTURER: _____          PILOT: _____

MODEL NUMBER: _____          SPOTTER: _____

PREFLIGHT CHECKLIST:

☐ Batteries Charged & Secure      ☐ Props OK & Tight              ☐ Compass Calibration
☐ Aircraft Hardware OK            ☐ Software / Firmware Update     ☐ Camera / FPV ON
☐ Equipment & Gear OK             ☐ Transmitter Control Power ON   ☐ Satellite Connection
☐ Transmitter Controls OK         ☐ Aircraft Power ON              ☐ Applications / Other Systems ON

OPERATIONAL CONDITIONS & PREFLIGHT NOTES:

| SESSION FLIGHT INTERVALS | FLIGHT TIMES | | | REMARKS |
| --- | --- | --- | --- | --- |
| | START | STOP | TOTAL | PROCEDURES & MANEUVERS |
| 1. | | | | |
| 2. | | | | |
| 3. | | | | |
| 4. | | | | |
| 5. | | | | |
| 6. | | | | |
| 7. | | | | |
| 8. | | | | |
| 9. | | | | |
| 10. | | | | |
| TOTAL HOURS FOR SESSION | | | | |
| TOTAL FORWARD | | | | |
| TOTAL TO DATE | | | | |

# II. FLIGHT MAP

Month _____ Day _____ Year _____

**POSTFLIGHT NOTES / JOURNAL ENTRIES:**

I CERTIFY THAT THE FOREGOING ENTRIES ARE TRUE AND CORRECT:

PILOT: _____ SPOTTER: _____

# I. FLIGHT ID / NO.: _____

FLIGHT LOCATION: _____     Weather: _____

UNMANNED AIRCRAFT SYSTEM:                      CREW:

MANUFACTURER: _____          PILOT: _____

MODEL NUMBER: _____          SPOTTER: _____

PREFLIGHT CHECKLIST:

☐ Batteries Charged & Secure    ☐ Props OK & Tight              ☐ Compass Calibration
☐ Aircraft Hardware OK          ☐ Software / Firmware Update     ☐ Camera / FPV ON
☐ Equipment & Gear OK           ☐ Transmitter Control Power ON   ☐ Satellite Connection
☐ Transmitter Controls OK       ☐ Aircraft Power ON              ☐ Applications / Other Systems ON

OPERATIONAL CONDITIONS & PREFLIGHT NOTES:

| SESSION FLIGHT INTERVALS | FLIGHT TIMES | | | REMARKS |
| --- | --- | --- | --- | --- |
| | START | STOP | TOTAL | PROCEDURES & MANEUVERS |
| 1. | | | | |
| 2. | | | | |
| 3. | | | | |
| 4. | | | | |
| 5. | | | | |
| 6. | | | | |
| 7. | | | | |
| 8. | | | | |
| 9. | | | | |
| 10. | | | | |
| TOTAL HOURS FOR SESSION | | | | |
| TOTAL FORWARD | | | | |
| TOTAL TO DATE | | | | |

## II. FLIGHT MAP

Month _____ Day _____ Year _____

**POSTFLIGHT NOTES / JOURNAL ENTRIES:**

I CERTIFY THAT THE FOREGOING ENTRIES ARE TRUE AND CORRECT:

PILOT: _____    SPOTTER: _____

# I. FLIGHT ID / NO.: _____

FLIGHT LOCATION: _____    Weather: _____

UNMANNED AIRCRAFT SYSTEM:    CREW:

MANUFACTURER: _____    PILOT: _____

MODEL NUMBER: _____    SPOTTER: _____

PREFLIGHT CHECKLIST:

☐ Batteries Charged & Secure    ☐ Props OK & Tight    ☐ Compass Calibration

☐ Aircraft Hardware OK    ☐ Software / Firmware Update    ☐ Camera / FPV ON

☐ Equipment & Gear OK    ☐ Transmitter Control Power ON    ☐ Satellite Connection

☐ Transmitter Controls OK    ☐ Aircraft Power ON    ☐ Applications / Other Systems ON

OPERATIONAL CONDITIONS & PREFLIGHT NOTES:

| SESSION FLIGHT INTERVALS | FLIGHT TIMES | | | REMARKS |
|---|---|---|---|---|
| | START | STOP | TOTAL | PROCEDURES & MANEUVERS |
| 1. | | | | |
| 2. | | | | |
| 3. | | | | |
| 4. | | | | |
| 5. | | | | |
| 6. | | | | |
| 7. | | | | |
| 8. | | | | |
| 9. | | | | |
| 10. | | | | |
| TOTAL HOURS FOR SESSION | | | | |
| TOTAL FORWARD | | | | |
| TOTAL TO DATE | | | | |

## II. FLIGHT MAP

Month _____ Day _____ Year _____

**POSTFLIGHT NOTES / JOURNAL ENTRIES:**

I CERTIFY THAT THE FOREGOING ENTRIES ARE TRUE AND CORRECT:

PILOT: _____ SPOTTER: _____

# I. FLIGHT ID / NO.: _____

FLIGHT LOCATION: _____     Weather: _____

UNMANNED AIRCRAFT SYSTEM:                         CREW:

MANUFACTURER: _____             PILOT: _____

MODEL NUMBER: _____             SPOTTER: _____

## PREFLIGHT CHECKLIST:

☐ Batteries Charged & Secure      ☐ Props OK & Tight                ☐ Compass Calibration
☐ Aircraft Hardware OK            ☐ Software / Firmware Update       ☐ Camera / FPV ON
☐ Equipment & Gear OK             ☐ Transmitter Control Power ON     ☐ Satellite Connection
☐ Transmitter Controls OK         ☐ Aircraft Power ON               ☐ Applications / Other Systems ON

## OPERATIONAL CONDITIONS & PREFLIGHT NOTES:

| SESSION FLIGHT INTERVALS | FLIGHT TIMES | | | REMARKS |
| --- | --- | --- | --- | --- |
| | START | STOP | TOTAL | PROCEDURES & MANEUVERS |
| 1. | | | | |
| 2. | | | | |
| 3. | | | | |
| 4. | | | | |
| 5. | | | | |
| 6. | | | | |
| 7. | | | | |
| 8. | | | | |
| 9. | | | | |
| 10. | | | | |
| TOTAL HOURS FOR SESSION | | | | |
| TOTAL FORWARD | | | | |
| TOTAL TO DATE | | | | |

# II. FLIGHT MAP

Month _____ Day _____ Year _____

POSTFLIGHT NOTES / JOURNAL ENTRIES:




I CERTIFY THAT THE FOREGOING ENTRIES ARE TRUE AND CORRECT:

PILOT: _____     SPOTTER: _____

27

# I. FLIGHT ID / NO.: _____

FLIGHT LOCATION: _____     Weather: _____

UNMANNED AIRCRAFT SYSTEM:                  CREW:

MANUFACTURER: _____        PILOT: _____

MODEL NUMBER: _____        SPOTTER: _____

PREFLIGHT CHECKLIST:

☐ Batteries Charged & Secure    ☐ Props OK & Tight              ☐ Compass Calibration
☐ Aircraft Hardware OK          ☐ Software / Firmware Update     ☐ Camera / FPV ON
☐ Equipment & Gear OK           ☐ Transmitter Control Power ON   ☐ Satellite Connection
☐ Transmitter Controls OK       ☐ Aircraft Power ON              ☐ Applications / Other Systems ON

OPERATIONAL CONDITIONS & PREFLIGHT NOTES:

| SESSION FLIGHT INTERVALS | FLIGHT TIMES | | | REMARKS |
| --- | --- | --- | --- | --- |
| | START | STOP | TOTAL | PROCEDURES & MANEUVERS |
| 1. | | | | |
| 2. | | | | |
| 3. | | | | |
| 4. | | | | |
| 5. | | | | |
| 6. | | | | |
| 7. | | | | |
| 8. | | | | |
| 9. | | | | |
| 10. | | | | |
| TOTAL HOURS FOR SESSION | | | | |
| TOTAL FORWARD | | | | |
| TOTAL TO DATE | | | | |

## II. FLIGHT MAP

Month _____ Day _____ Year _____

**POSTFLIGHT NOTES / JOURNAL ENTRIES:**

I CERTIFY THAT THE FOREGOING ENTRIES ARE TRUE AND CORRECT:

PILOT: _____ SPOTTER: _____

# I. FLIGHT ID / NO.: _____

FLIGHT LOCATION: _____    Weather: _____

UNMANNED AIRCRAFT SYSTEM:                    CREW:

MANUFACTURER: _____       PILOT: _____

MODEL NUMBER: _____       SPOTTER: _____

PREFLIGHT CHECKLIST:

☐ Batteries Charged & Secure   ☐ Props OK & Tight              ☐ Compass Calibration
☐ Aircraft Hardware OK         ☐ Software / Firmware Update     ☐ Camera / FPV ON
☐ Equipment & Gear OK          ☐ Transmitter Control Power ON   ☐ Satellite Connection
☐ Transmitter Controls OK      ☐ Aircraft Power ON              ☐ Applications / Other Systems ON

OPERATIONAL CONDITIONS & PREFLIGHT NOTES:

| SESSION FLIGHT INTERVALS | FLIGHT TIMES | | | REMARKS |
|---|---|---|---|---|
| | START | STOP | TOTAL | PROCEDURES & MANEUVERS |
| 1. | | | | |
| 2. | | | | |
| 3. | | | | |
| 4. | | | | |
| 5. | | | | |
| 6. | | | | |
| 7. | | | | |
| 8. | | | | |
| 9. | | | | |
| 10. | | | | |
| TOTAL HOURS FOR SESSION | | | | |
| TOTAL FORWARD | | | | |
| TOTAL TO DATE | | | | |

## II. FLIGHT MAP

**POSTFLIGHT NOTES / JOURNAL ENTRIES:**

I CERTIFY THAT THE FOREGOING ENTRIES ARE TRUE AND CORRECT:

PILOT: _____ SPOTTER: _____

# I. FLIGHT ID / NO.: _____

FLIGHT LOCATION: _____    Weather: _____

UNMANNED AIRCRAFT SYSTEM:          CREW:

MANUFACTURER: _____    PILOT: _____

MODEL NUMBER: _____    SPOTTER: _____

PREFLIGHT CHECKLIST:

☐ Batteries Charged & Secure   ☐ Props OK & Tight            ☐ Compass Calibration
☐ Aircraft Hardware OK          ☐ Software / Firmware Update   ☐ Camera / FPV ON
☐ Equipment & Gear OK           ☐ Transmitter Control Power ON ☐ Satellite Connection
☐ Transmitter Controls OK       ☐ Aircraft Power ON           ☐ Applications / Other Systems ON

OPERATIONAL CONDITIONS & PREFLIGHT NOTES:

| SESSION FLIGHT INTERVALS | FLIGHT TIMES | | | REMARKS |
| --- | --- | --- | --- | --- |
| | START | STOP | TOTAL | PROCEDURES & MANEUVERS |
| 1. | | | | |
| 2. | | | | |
| 3. | | | | |
| 4. | | | | |
| 5. | | | | |
| 6. | | | | |
| 7. | | | | |
| 8. | | | | |
| 9. | | | | |
| 10. | | | | |
| TOTAL HOURS FOR SESSION | | | | |
| TOTAL FORWARD | | | | |
| TOTAL TO DATE | | | | |

## II. FLIGHT MAP

Month _____ Day _____ Year _____

**POSTFLIGHT NOTES / JOURNAL ENTRIES:**

I CERTIFY THAT THE FOREGOING ENTRIES ARE TRUE AND CORRECT:

PILOT: _____     SPOTTER: _____

# I. FLIGHT ID / NO.: _____

FLIGHT LOCATION: _____     Weather: _____

UNMANNED AIRCRAFT SYSTEM:                   CREW:

MANUFACTURER: _____         PILOT: _____

MODEL NUMBER: _____         SPOTTER: _____

PREFLIGHT CHECKLIST:

☐ Batteries Charged & Secure     ☐ Props OK & Tight              ☐ Compass Calibration
☐ Aircraft Hardware OK           ☐ Software / Firmware Update     ☐ Camera / FPV ON
☐ Equipment & Gear OK            ☐ Transmitter Control Power ON   ☐ Satellite Connection
☐ Transmitter Controls OK        ☐ Aircraft Power ON              ☐ Applications / Other Systems ON

OPERATIONAL CONDITIONS & PREFLIGHT NOTES:

| SESSION FLIGHT INTERVALS | FLIGHT TIMES | | | REMARKS |
| --- | --- | --- | --- | --- |
| | START | STOP | TOTAL | PROCEDURES & MANEUVERS |
| 1. | | | | |
| 2. | | | | |
| 3. | | | | |
| 4. | | | | |
| 5. | | | | |
| 6. | | | | |
| 7. | | | | |
| 8. | | | | |
| 9. | | | | |
| 10. | | | | |
| TOTAL HOURS FOR SESSION | | | | |
| TOTAL FORWARD | | | | |
| TOTAL TO DATE | | | | |

## II. FLIGHT MAP

Month _____ Day _____ Year _____

POSTFLIGHT NOTES / JOURNAL ENTRIES:

I CERTIFY THAT THE FOREGOING ENTRIES ARE TRUE AND CORRECT:

PILOT: _____ SPOTTER: _____

# I. FLIGHT ID / NO.: _____

FLIGHT LOCATION: _____  Weather: _____

## UNMANNED AIRCRAFT SYSTEM:                    CREW:

MANUFACTURER: _____       PILOT: _____

MODEL NUMBER: _____       SPOTTER: _____

## PREFLIGHT CHECKLIST:

☐ Batteries Charged & Secure    ☐ Props OK & Tight              ☐ Compass Calibration

☐ Aircraft Hardware OK          ☐ Software / Firmware Update     ☐ Camera / FPV ON

☐ Equipment & Gear OK           ☐ Transmitter Control Power ON   ☐ Satellite Connection

☐ Transmitter Controls OK       ☐ Aircraft Power ON              ☐ Applications / Other Systems ON

## OPERATIONAL CONDITIONS & PREFLIGHT NOTES:

| SESSION FLIGHT INTERVALS | FLIGHT TIMES | | | REMARKS |
|---|---|---|---|---|
| | START | STOP | TOTAL | PROCEDURES & MANEUVERS |
| 1. | | | | |
| 2. | | | | |
| 3. | | | | |
| 4. | | | | |
| 5. | | | | |
| 6. | | | | |
| 7. | | | | |
| 8. | | | | |
| 9. | | | | |
| 10. | | | | |
| TOTAL HOURS FOR SESSION | | | | |
| TOTAL FORWARD | | | | |
| TOTAL TO DATE | | | | |

## II. FLIGHT MAP

Month _____ Day _____ Year _____

POSTFLIGHT NOTES / JOURNAL ENTRIES:

I CERTIFY THAT THE FOREGOING ENTRIES ARE TRUE AND CORRECT:

PILOT: _____  SPOTTER: _____

# I. FLIGHT ID / NO.: _____

FLIGHT LOCATION: _____     Weather: _____

UNMANNED AIRCRAFT SYSTEM:                    CREW:

MANUFACTURER: _____         PILOT: _____

MODEL NUMBER: _____         SPOTTER: _____

PREFLIGHT CHECKLIST:

| | | |
|---|---|---|
| ☐ Batteries Charged & Secure | ☐ Props OK & Tight | ☐ Compass Calibration |
| ☐ Aircraft Hardware OK | ☐ Software / Firmware Update | ☐ Camera / FPV ON |
| ☐ Equipment & Gear OK | ☐ Transmitter Control Power ON | ☐ Satellite Connection |
| ☐ Transmitter Controls OK | ☐ Aircraft Power ON | ☐ Applications / Other Systems ON |

OPERATIONAL CONDITIONS & PREFLIGHT NOTES:

| SESSION FLIGHT INTERVALS | FLIGHT TIMES | | | REMARKS |
|---|---|---|---|---|
| | START | STOP | TOTAL | PROCEDURES & MANEUVERS |
| 1. | | | | |
| 2. | | | | |
| 3. | | | | |
| 4. | | | | |
| 5. | | | | |
| 6. | | | | |
| 7. | | | | |
| 8. | | | | |
| 9. | | | | |
| 10. | | | | |
| TOTAL HOURS FOR SESSION | | | | |
| TOTAL FORWARD | | | | |
| TOTAL TO DATE | | | | |

## II. FLIGHT MAP

Month _____ Day _____ Year _____

**POSTFLIGHT NOTES / JOURNAL ENTRIES:**

I CERTIFY THAT THE FOREGOING ENTRIES ARE TRUE AND CORRECT:

PILOT: _____     SPOTTER: _____

# I. FLIGHT ID / NO.: _____

FLIGHT LOCATION: _____     Weather: _____

UNMANNED AIRCRAFT SYSTEM:                        CREW:

MANUFACTURER: _____           PILOT: _____

MODEL NUMBER: _____           SPOTTER: _____

PREFLIGHT CHECKLIST:

☐ Batteries Charged & Secure      ☐ Props OK & Tight                ☐ Compass Calibration
☐ Aircraft Hardware OK            ☐ Software / Firmware Update      ☐ Camera / FPV ON
☐ Equipment & Gear OK             ☐ Transmitter Control Power ON    ☐ Satellite Connection
☐ Transmitter Controls OK         ☐ Aircraft Power ON               ☐ Applications / Other Systems ON

OPERATIONAL CONDITIONS & PREFLIGHT NOTES:

| SESSION FLIGHT INTERVALS | FLIGHT TIMES | | | REMARKS |
| --- | --- | --- | --- | --- |
| | START | STOP | TOTAL | PROCEDURES & MANEUVERS |
| 1. | | | | |
| 2. | | | | |
| 3. | | | | |
| 4. | | | | |
| 5. | | | | |
| 6. | | | | |
| 7. | | | | |
| 8. | | | | |
| 9. | | | | |
| 10. | | | | |
| TOTAL HOURS FOR SESSION | | | | |
| TOTAL FORWARD | | | | |
| TOTAL TO DATE | | | | |

# II. FLIGHT MAP

Month _____ Day _____ Year _____

**POSTFLIGHT NOTES / JOURNAL ENTRIES:**

I CERTIFY THAT THE FOREGOING ENTRIES ARE TRUE AND CORRECT:

PILOT: _____ SPOTTER: _____

# I. FLIGHT ID / NO.: _____

FLIGHT LOCATION: _____    Weather: _____

UNMANNED AIRCRAFT SYSTEM:                    CREW:

MANUFACTURER: _____        PILOT: _____

MODEL NUMBER: _____        SPOTTER: _____

PREFLIGHT CHECKLIST:

☐ Batteries Charged & Secure    ☐ Props OK & Tight               ☐ Compass Calibration
☐ Aircraft Hardware OK          ☐ Software / Firmware Update      ☐ Camera / FPV ON
☐ Equipment & Gear OK           ☐ Transmitter Control Power ON    ☐ Satellite Connection
☐ Transmitter Controls OK       ☐ Aircraft Power ON               ☐ Applications / Other Systems ON

OPERATIONAL CONDITIONS & PREFLIGHT NOTES:

| SESSION FLIGHT INTERVALS | FLIGHT TIMES | | | REMARKS |
| --- | --- | --- | --- | --- |
| | START | STOP | TOTAL | PROCEDURES & MANEUVERS |
| 1. | | | | |
| 2. | | | | |
| 3. | | | | |
| 4. | | | | |
| 5. | | | | |
| 6. | | | | |
| 7. | | | | |
| 8. | | | | |
| 9. | | | | |
| 10. | | | | |
| TOTAL HOURS FOR SESSION | | | | |
| TOTAL FORWARD | | | | |
| TOTAL TO DATE | | | | |

## II. FLIGHT MAP

Month _____ Day _____ Year _____

**POSTFLIGHT NOTES / JOURNAL ENTRIES:**

I CERTIFY THAT THE FOREGOING ENTRIES ARE TRUE AND CORRECT:

PILOT: _____ SPOTTER: _____

# I. FLIGHT ID / NO.: _____

FLIGHT LOCATION: _____    Weather: _____

UNMANNED AIRCRAFT SYSTEM:                CREW:

MANUFACTURER: _____       PILOT: _____

MODEL NUMBER: _____       SPOTTER: _____

PREFLIGHT CHECKLIST:

☐ Batteries Charged & Secure    ☐ Props OK & Tight              ☐ Compass Calibration
☐ Aircraft Hardware OK          ☐ Software / Firmware Update     ☐ Camera / FPV ON
☐ Equipment & Gear OK           ☐ Transmitter Control Power ON   ☐ Satellite Connection
☐ Transmitter Controls OK       ☐ Aircraft Power ON              ☐ Applications / Other Systems ON

OPERATIONAL CONDITIONS & PREFLIGHT NOTES:

| SESSION FLIGHT INTERVALS | FLIGHT TIMES | | | REMARKS |
|---|---|---|---|---|
| | START | STOP | TOTAL | PROCEDURES & MANEUVERS |
| 1. | | | | |
| 2. | | | | |
| 3. | | | | |
| 4. | | | | |
| 5. | | | | |
| 6. | | | | |
| 7. | | | | |
| 8. | | | | |
| 9. | | | | |
| 10. | | | | |
| TOTAL HOURS FOR SESSION | | | | |
| TOTAL FORWARD | | | | |
| TOTAL TO DATE | | | | |

## II. FLIGHT MAP

Month _____ Day _____ Year _____

**POSTFLIGHT NOTES / JOURNAL ENTRIES:**

I CERTIFY THAT THE FOREGOING ENTRIES ARE TRUE AND CORRECT:

PILOT: _____   SPOTTER: _____

# I. FLIGHT ID / NO.: _____

FLIGHT LOCATION: _____     Weather: _____

UNMANNED AIRCRAFT SYSTEM:                   CREW:

MANUFACTURER: _____         PILOT: _____

MODEL NUMBER: _____         SPOTTER: _____

## PREFLIGHT CHECKLIST:

☐  Batteries Charged & Secure      ☐  Props OK & Tight              ☐  Compass Calibration
☐  Aircraft Hardware OK            ☐  Software / Firmware Update     ☐  Camera / FPV ON
☐  Equipment & Gear OK             ☐  Transmitter Control Power ON   ☐  Satellite Connection
☐  Transmitter Controls OK         ☐  Aircraft Power ON              ☐  Applications / Other Systems ON

## OPERATIONAL CONDITIONS & PREFLIGHT NOTES:

| SESSION FLIGHT INTERVALS | FLIGHT TIMES | | | REMARKS |
|---|---|---|---|---|
| | START | STOP | TOTAL | PROCEDURES & MANEUVERS |
| 1. | | | | |
| 2. | | | | |
| 3. | | | | |
| 4. | | | | |
| 5. | | | | |
| 6. | | | | |
| 7. | | | | |
| 8. | | | | |
| 9. | | | | |
| 10. | | | | |
| TOTAL HOURS FOR SESSION | | | | |
| TOTAL FORWARD | | | | |
| TOTAL TO DATE | | | | |

# II. FLIGHT MAP

Month _____ Day _____ Year _____

**POSTFLIGHT NOTES / JOURNAL ENTRIES:**

I CERTIFY THAT THE FOREGOING ENTRIES ARE TRUE AND CORRECT:

PILOT: _____ SPOTTER: _____

# I. FLIGHT ID / NO.: _____

FLIGHT LOCATION: _____     Weather: _____

UNMANNED AIRCRAFT SYSTEM:                    CREW:

MANUFACTURER: _____          PILOT: _____

MODEL NUMBER: _____          SPOTTER: _____

## PREFLIGHT CHECKLIST:

☐ Batteries Charged & Secure    ☐ Props OK & Tight              ☐ Compass Calibration
☐ Aircraft Hardware OK          ☐ Software / Firmware Update     ☐ Camera / FPV ON
☐ Equipment & Gear OK           ☐ Transmitter Control Power ON   ☐ Satellite Connection
☐ Transmitter Controls OK       ☐ Aircraft Power ON              ☐ Applications / Other Systems ON

## OPERATIONAL CONDITIONS & PREFLIGHT NOTES:

| SESSION FLIGHT INTERVALS | FLIGHT TIMES | | | REMARKS |
| --- | --- | --- | --- | --- |
| | START | STOP | TOTAL | PROCEDURES & MANEUVERS |
| 1. | | | | |
| 2. | | | | |
| 3. | | | | |
| 4. | | | | |
| 5. | | | | |
| 6. | | | | |
| 7. | | | | |
| 8. | | | | |
| 9. | | | | |
| 10. | | | | |
| TOTAL HOURS FOR SESSION | | | | |
| TOTAL FORWARD | | | | |
| TOTAL TO DATE | | | | |

## II. FLIGHT MAP

Month _____ Day _____ Year _____

POSTFLIGHT NOTES / JOURNAL ENTRIES:

I CERTIFY THAT THE FOREGOING ENTRIES ARE TRUE AND CORRECT:

PILOT: _____ SPOTTER: _____

# I. FLIGHT ID / NO.: _____

FLIGHT LOCATION: _____    Weather: _____

UNMANNED AIRCRAFT SYSTEM:                    CREW:

MANUFACTURER: _____         PILOT: _____

MODEL NUMBER: _____          SPOTTER: _____

PREFLIGHT CHECKLIST:

☐ Batteries Charged & Secure      ☐ Props OK & Tight              ☐ Compass Calibration
☐ Aircraft Hardware OK            ☐ Software / Firmware Update     ☐ Camera / FPV ON
☐ Equipment & Gear OK             ☐ Transmitter Control Power ON   ☐ Satellite Connection
☐ Transmitter Controls OK         ☐ Aircraft Power ON              ☐ Applications / Other Systems ON

OPERATIONAL CONDITIONS & PREFLIGHT NOTES:

| SESSION FLIGHT INTERVALS | FLIGHT TIMES | | | REMARKS |
| --- | --- | --- | --- | --- |
| | START | STOP | TOTAL | PROCEDURES & MANEUVERS |
| 1. | | | | |
| 2. | | | | |
| 3. | | | | |
| 4. | | | | |
| 5. | | | | |
| 6. | | | | |
| 7. | | | | |
| 8. | | | | |
| 9. | | | | |
| 10. | | | | |
| TOTAL HOURS FOR SESSION | | | | |
| TOTAL FORWARD | | | | |
| TOTAL TO DATE | | | | |

## II. FLIGHT MAP

Month _____ Day _____ Year _____

POSTFLIGHT NOTES / JOURNAL ENTRIES:

I CERTIFY THAT THE FOREGOING ENTRIES ARE TRUE AND CORRECT:

PILOT: _____     SPOTTER: _____

# I. FLIGHT ID / NO.: _____

FLIGHT LOCATION: _____          Weather: _____

UNMANNED AIRCRAFT SYSTEM:                         CREW:

MANUFACTURER: _____              PILOT: _____

MODEL NUMBER: _____              SPOTTER: _____

PREFLIGHT CHECKLIST:

☐ Batteries Charged & Secure      ☐ Props OK & Tight              ☐ Compass Calibration
☐ Aircraft Hardware OK            ☐ Software / Firmware Update     ☐ Camera / FPV ON
☐ Equipment & Gear OK             ☐ Transmitter Control Power ON   ☐ Satellite Connection
☐ Transmitter Controls OK         ☐ Aircraft Power ON              ☐ Applications / Other Systems ON

OPERATIONAL CONDITIONS & PREFLIGHT NOTES:

| SESSION FLIGHT INTERVALS | FLIGHT TIMES | | | REMARKS |
|---|---|---|---|---|
| | START | STOP | TOTAL | PROCEDURES & MANEUVERS |
| 1. | | | | |
| 2. | | | | |
| 3. | | | | |
| 4. | | | | |
| 5. | | | | |
| 6. | | | | |
| 7. | | | | |
| 8. | | | | |
| 9. | | | | |
| 10. | | | | |
| TOTAL HOURS FOR SESSION | | | | |
| TOTAL FORWARD | | | | |
| TOTAL TO DATE | | | | |

## II. FLIGHT MAP

Month _____ Day _____ Year _____

**POSTFLIGHT NOTES / JOURNAL ENTRIES:**

I CERTIFY THAT THE FOREGOING ENTRIES ARE TRUE AND CORRECT:

PILOT: _____ SPOTTER: _____

# I. FLIGHT ID / NO.: _____

FLIGHT LOCATION: _____     Weather: _____

UNMANNED AIRCRAFT SYSTEM:                    CREW:

MANUFACTURER: _____     PILOT: _____

MODEL NUMBER: _____     SPOTTER: _____

PREFLIGHT CHECKLIST:

☐ Batteries Charged & Secure      ☐ Props OK & Tight                ☐ Compass Calibration
☐ Aircraft Hardware OK            ☐ Software / Firmware Update       ☐ Camera / FPV ON
☐ Equipment & Gear OK             ☐ Transmitter Control Power ON     ☐ Satellite Connection
☐ Transmitter Controls OK         ☐ Aircraft Power ON                ☐ Applications / Other Systems ON

OPERATIONAL CONDITIONS & PREFLIGHT NOTES:

| SESSION FLIGHT INTERVALS | FLIGHT TIMES | | | REMARKS |
| --- | --- | --- | --- | --- |
| | START | STOP | TOTAL | PROCEDURES & MANEUVERS |
| 1. | | | | |
| 2. | | | | |
| 3. | | | | |
| 4. | | | | |
| 5. | | | | |
| 6. | | | | |
| 7. | | | | |
| 8. | | | | |
| 9. | | | | |
| 10. | | | | |
| TOTAL HOURS FOR SESSION | | | | |
| TOTAL FORWARD | | | | |
| TOTAL TO DATE | | | | |

## II. FLIGHT MAP

Month _____ Day _____ Year _____

**POSTFLIGHT NOTES / JOURNAL ENTRIES:**

I CERTIFY THAT THE FOREGOING ENTRIES ARE TRUE AND CORRECT:

PILOT: _____  SPOTTER: _____

# I. FLIGHT ID / NO.: _____

FLIGHT LOCATION: _____     Weather: _____

UNMANNED AIRCRAFT SYSTEM:                          CREW:

MANUFACTURER: _____         PILOT: _____

MODEL NUMBER: _____         SPOTTER: _____

PREFLIGHT CHECKLIST:

☐ Batteries Charged & Secure      ☐ Props OK & Tight               ☐ Compass Calibration

☐ Aircraft Hardware OK            ☐ Software / Firmware Update      ☐ Camera / FPV ON

☐ Equipment & Gear OK             ☐ Transmitter Control Power ON    ☐ Satellite Connection

☐ Transmitter Controls OK         ☐ Aircraft Power ON               ☐ Applications / Other Systems ON

OPERATIONAL CONDITIONS & PREFLIGHT NOTES:

| SESSION FLIGHT INTERVALS | FLIGHT TIMES | | | REMARKS |
|---|---|---|---|---|
| | START | STOP | TOTAL | PROCEDURES & MANEUVERS |
| 1. | | | | |
| 2. | | | | |
| 3. | | | | |
| 4. | | | | |
| 5. | | | | |
| 6. | | | | |
| 7. | | | | |
| 8. | | | | |
| 9. | | | | |
| 10. | | | | |
| TOTAL HOURS FOR SESSION | | | | |
| TOTAL FORWARD | | | | |
| TOTAL TO DATE | | | | |

## II. FLIGHT MAP

Month _____ Day _____ Year _____

POSTFLIGHT NOTES / JOURNAL ENTRIES:

I CERTIFY THAT THE FOREGOING ENTRIES ARE TRUE AND CORRECT:

PILOT: _____     SPOTTER: _____

# I. FLIGHT ID / NO.: _____

FLIGHT LOCATION: _____     Weather: _____

UNMANNED AIRCRAFT SYSTEM:                          CREW:

MANUFACTURER: _____              PILOT: _____

MODEL NUMBER: _____              SPOTTER: _____

PREFLIGHT CHECKLIST:

☐ Batteries Charged & Secure    ☐ Props OK & Tight              ☐ Compass Calibration
☐ Aircraft Hardware OK          ☐ Software / Firmware Update    ☐ Camera / FPV ON
☐ Equipment & Gear OK           ☐ Transmitter Control Power ON  ☐ Satellite Connection
☐ Transmitter Controls OK       ☐ Aircraft Power ON             ☐ Applications / Other Systems ON

OPERATIONAL CONDITIONS & PREFLIGHT NOTES:

| SESSION FLIGHT INTERVALS | FLIGHT TIMES | | | REMARKS |
| --- | --- | --- | --- | --- |
| | START | STOP | TOTAL | PROCEDURES & MANEUVERS |
| 1. | | | | |
| 2. | | | | |
| 3. | | | | |
| 4. | | | | |
| 5. | | | | |
| 6. | | | | |
| 7. | | | | |
| 8. | | | | |
| 9. | | | | |
| 10. | | | | |
| TOTAL HOURS FOR SESSION | | | | |
| TOTAL FORWARD | | | | |
| TOTAL TO DATE | | | | |

## II. FLIGHT MAP

Month _____ Day _____ Year _____

POSTFLIGHT NOTES / JOURNAL ENTRIES:

I CERTIFY THAT THE FOREGOING ENTRIES ARE TRUE AND CORRECT:

PILOT: _____ SPOTTER: _____

# I. FLIGHT ID / NO.: _____

FLIGHT LOCATION: _____     Weather: _____

UNMANNED AIRCRAFT SYSTEM:                   CREW:

MANUFACTURER: _____         PILOT: _____

MODEL NUMBER: _____          SPOTTER: _____

## PREFLIGHT CHECKLIST:

☐ Batteries Charged & Secure     ☐ Props OK & Tight              ☐ Compass Calibration

☐ Aircraft Hardware OK           ☐ Software / Firmware Update     ☐ Camera / FPV ON

☐ Equipment & Gear OK            ☐ Transmitter Control Power ON   ☐ Satellite Connection

☐ Transmitter Controls OK        ☐ Aircraft Power ON              ☐ Applications / Other Systems ON

## OPERATIONAL CONDITIONS & PREFLIGHT NOTES:

| SESSION FLIGHT INTERVALS | FLIGHT TIMES | | | REMARKS |
| --- | --- | --- | --- | --- |
| | START | STOP | TOTAL | PROCEDURES & MANEUVERS |
| 1. | | | | |
| 2. | | | | |
| 3. | | | | |
| 4. | | | | |
| 5. | | | | |
| 6. | | | | |
| 7. | | | | |
| 8. | | | | |
| 9. | | | | |
| 10. | | | | |
| TOTAL HOURS FOR SESSION | | | | |
| TOTAL FORWARD | | | | |
| TOTAL TO DATE | | | | |

# II. FLIGHT MAP

Month _____ Day _____ Year _____

POSTFLIGHT NOTES / JOURNAL ENTRIES:

I CERTIFY THAT THE FOREGOING ENTRIES ARE TRUE AND CORRECT:

PILOT: _____  SPOTTER: _____

# I. FLIGHT ID / NO.: _____

FLIGHT LOCATION: _____     Weather: _____

UNMANNED AIRCRAFT SYSTEM:                    CREW:

MANUFACTURER: _____        PILOT: _____

MODEL NUMBER: _____        SPOTTER: _____

PREFLIGHT CHECKLIST:

☐ Batteries Charged & Secure    ☐ Props OK & Tight              ☐ Compass Calibration
☐ Aircraft Hardware OK          ☐ Software / Firmware Update    ☐ Camera / FPV ON
☐ Equipment & Gear OK           ☐ Transmitter Control Power ON  ☐ Satellite Connection
☐ Transmitter Controls OK       ☐ Aircraft Power ON             ☐ Applications / Other Systems ON

OPERATIONAL CONDITIONS & PREFLIGHT NOTES:

| SESSION FLIGHT INTERVALS | FLIGHT TIMES | | | REMARKS |
| --- | --- | --- | --- | --- |
| | START | STOP | TOTAL | PROCEDURES & MANEUVERS |
| 1. | | | | |
| 2. | | | | |
| 3. | | | | |
| 4. | | | | |
| 5. | | | | |
| 6. | | | | |
| 7. | | | | |
| 8. | | | | |
| 9. | | | | |
| 10. | | | | |
| TOTAL HOURS FOR SESSION | | | | |
| TOTAL FORWARD | | | | |
| TOTAL TO DATE | | | | |

## II. FLIGHT MAP

Month _____ Day _____ Year _____

**POSTFLIGHT NOTES / JOURNAL ENTRIES:**

I CERTIFY THAT THE FOREGOING ENTRIES ARE TRUE AND CORRECT:

PILOT: _____ SPOTTER: _____

# I. FLIGHT ID / NO.: _____

FLIGHT LOCATION: _____    Weather: _____

UNMANNED AIRCRAFT SYSTEM:                  CREW:

MANUFACTURER: _____      PILOT: _____

MODEL NUMBER: _____      SPOTTER: _____

## PREFLIGHT CHECKLIST:

☐ Batteries Charged & Secure      ☐ Props OK & Tight                ☐ Compass Calibration
☐ Aircraft Hardware OK            ☐ Software / Firmware Update       ☐ Camera / FPV ON
☐ Equipment & Gear OK             ☐ Transmitter Control Power ON     ☐ Satellite Connection
☐ Transmitter Controls OK         ☐ Aircraft Power ON                ☐ Applications / Other Systems ON

## OPERATIONAL CONDITIONS & PREFLIGHT NOTES:

| SESSION FLIGHT INTERVALS | FLIGHT TIMES | | | REMARKS |
| --- | --- | --- | --- | --- |
| | START | STOP | TOTAL | PROCEDURES & MANEUVERS |
| 1. | | | | |
| 2. | | | | |
| 3. | | | | |
| 4. | | | | |
| 5. | | | | |
| 6. | | | | |
| 7. | | | | |
| 8. | | | | |
| 9. | | | | |
| 10. | | | | |
| TOTAL HOURS FOR SESSION | | | | |
| TOTAL FORWARD | | | | |
| TOTAL TO DATE | | | | |

# II. FLIGHT MAP

Month _____ Day _____ Year _____

**POSTFLIGHT NOTES / JOURNAL ENTRIES:**

I CERTIFY THAT THE FOREGOING ENTRIES ARE TRUE AND CORRECT:

PILOT: _____     SPOTTER: _____

# I. FLIGHT ID / NO.: _____

FLIGHT LOCATION: _____     Weather: _____

UNMANNED AIRCRAFT SYSTEM:                      CREW:

MANUFACTURER: _____          PILOT: _____

MODEL NUMBER: _____          SPOTTER: _____

PREFLIGHT CHECKLIST:

☐ Batteries Charged & Secure     ☐ Props OK & Tight                ☐ Compass Calibration
☐ Aircraft Hardware OK           ☐ Software / Firmware Update       ☐ Camera / FPV ON
☐ Equipment & Gear OK            ☐ Transmitter Control Power ON     ☐ Satellite Connection
☐ Transmitter Controls OK        ☐ Aircraft Power ON               ☐ Applications / Other Systems ON

OPERATIONAL CONDITIONS & PREFLIGHT NOTES:

| SESSION FLIGHT INTERVALS | FLIGHT TIMES | | | REMARKS |
| --- | --- | --- | --- | --- |
| | START | STOP | TOTAL | PROCEDURES & MANEUVERS |
| 1. | | | | |
| 2. | | | | |
| 3. | | | | |
| 4. | | | | |
| 5. | | | | |
| 6. | | | | |
| 7. | | | | |
| 8. | | | | |
| 9. | | | | |
| 10. | | | | |
| TOTAL HOURS FOR SESSION | | | | |
| TOTAL FORWARD | | | | |
| TOTAL TO DATE | | | | |

## II. FLIGHT MAP

Month _____ Day _____ Year _____

POSTFLIGHT NOTES / JOURNAL ENTRIES:

I CERTIFY THAT THE FOREGOING ENTRIES ARE TRUE AND CORRECT:

PILOT: _____     SPOTTER: _____

# I. FLIGHT ID / NO.: _____

FLIGHT LOCATION: _____     Weather: _____

## UNMANNED AIRCRAFT SYSTEM:                CREW:

MANUFACTURER: _____        PILOT: _____

MODEL NUMBER: _____        SPOTTER: _____

## PREFLIGHT CHECKLIST:

☐ Batteries Charged & Secure      ☐ Props OK & Tight              ☐ Compass Calibration
☐ Aircraft Hardware OK            ☐ Software / Firmware Update     ☐ Camera / FPV ON
☐ Equipment & Gear OK             ☐ Transmitter Control Power ON   ☐ Satellite Connection
☐ Transmitter Controls OK         ☐ Aircraft Power ON              ☐ Applications / Other Systems ON

## OPERATIONAL CONDITIONS & PREFLIGHT NOTES:

| SESSION FLIGHT INTERVALS | FLIGHT TIMES | | | REMARKS |
| --- | --- | --- | --- | --- |
| | START | STOP | TOTAL | PROCEDURES & MANEUVERS |
| 1. | | | | |
| 2. | | | | |
| 3. | | | | |
| 4. | | | | |
| 5. | | | | |
| 6. | | | | |
| 7. | | | | |
| 8. | | | | |
| 9. | | | | |
| 10. | | | | |
| TOTAL HOURS FOR SESSION | | | | |
| TOTAL FORWARD | | | | |
| TOTAL TO DATE | | | | |

# II. FLIGHT MAP

Month _____ Day _____ Year _____

**POSTFLIGHT NOTES / JOURNAL ENTRIES:**

I CERTIFY THAT THE FOREGOING ENTRIES ARE TRUE AND CORRECT:

PILOT: _____    SPOTTER: _____

# I. FLIGHT ID / NO.: _____

FLIGHT LOCATION: _____     Weather: _____

UNMANNED AIRCRAFT SYSTEM:                  CREW:

MANUFACTURER: _____        PILOT: _____

MODEL NUMBER: _____        SPOTTER: _____

PREFLIGHT CHECKLIST:

☐ Batteries Charged & Secure     ☐ Props OK & Tight              ☐ Compass Calibration
☐ Aircraft Hardware OK           ☐ Software / Firmware Update     ☐ Camera / FPV ON
☐ Equipment & Gear OK            ☐ Transmitter Control Power ON   ☐ Satellite Connection
☐ Transmitter Controls OK        ☐ Aircraft Power ON              ☐ Applications / Other Systems ON

OPERATIONAL CONDITIONS & PREFLIGHT NOTES:

| SESSION FLIGHT INTERVALS | FLIGHT TIMES | | | REMARKS |
| --- | --- | --- | --- | --- |
| | START | STOP | TOTAL | PROCEDURES & MANEUVERS |
| 1. | | | | |
| 2. | | | | |
| 3. | | | | |
| 4. | | | | |
| 5. | | | | |
| 6. | | | | |
| 7. | | | | |
| 8. | | | | |
| 9. | | | | |
| 10. | | | | |
| TOTAL HOURS FOR SESSION | | | | |
| TOTAL FORWARD | | | | |
| TOTAL TO DATE | | | | |

## II. FLIGHT MAP

Month _____ Day _____ Year _____

POSTFLIGHT NOTES / JOURNAL ENTRIES:

I CERTIFY THAT THE FOREGOING ENTRIES ARE TRUE AND CORRECT:

PILOT: _____ SPOTTER: _____

# I. FLIGHT ID / NO.: _____

FLIGHT LOCATION: _____     Weather: _____

UNMANNED AIRCRAFT SYSTEM:                    CREW:

MANUFACTURER: _____          PILOT: _____

MODEL NUMBER: _____          SPOTTER: _____

PREFLIGHT CHECKLIST:

☐ Batteries Charged & Secure      ☐ Props OK & Tight            ☐ Compass Calibration
☐ Aircraft Hardware OK            ☐ Software / Firmware Update   ☐ Camera / FPV ON
☐ Equipment & Gear OK             ☐ Transmitter Control Power ON ☐ Satellite Connection
☐ Transmitter Controls OK         ☐ Aircraft Power ON           ☐ Applications / Other Systems ON

OPERATIONAL CONDITIONS & PREFLIGHT NOTES:

| SESSION FLIGHT INTERVALS | FLIGHT TIMES | | | REMARKS |
| --- | --- | --- | --- | --- |
| | START | STOP | TOTAL | PROCEDURES & MANEUVERS |
| 1. | | | | |
| 2. | | | | |
| 3. | | | | |
| 4. | | | | |
| 5. | | | | |
| 6. | | | | |
| 7. | | | | |
| 8. | | | | |
| 9. | | | | |
| 10. | | | | |
| TOTAL HOURS FOR SESSION | | | | |
| TOTAL FORWARD | | | | |
| TOTAL TO DATE | | | | |

## II. FLIGHT MAP

Month _____ Day _____ Year _____

POSTFLIGHT NOTES / JOURNAL ENTRIES:

I CERTIFY THAT THE FOREGOING ENTRIES ARE TRUE AND CORRECT:

PILOT: _____ SPOTTER: _____

# I. FLIGHT ID / NO.: _____

FLIGHT LOCATION: _____   Weather: _____

UNMANNED AIRCRAFT SYSTEM:                         CREW:

MANUFACTURER: _____            PILOT: _____

MODEL NUMBER: _____            SPOTTER: _____

PREFLIGHT CHECKLIST:

☐ Batteries Charged & Secure    ☐ Props OK & Tight              ☐ Compass Calibration
☐ Aircraft Hardware OK          ☐ Software / Firmware Update    ☐ Camera / FPV ON
☐ Equipment & Gear OK           ☐ Transmitter Control Power ON  ☐ Satellite Connection
☐ Transmitter Controls OK       ☐ Aircraft Power ON             ☐ Applications / Other Systems ON

OPERATIONAL CONDITIONS & PREFLIGHT NOTES:

| SESSION FLIGHT INTERVALS | FLIGHT TIMES | | | REMARKS |
| --- | --- | --- | --- | --- |
| | START | STOP | TOTAL | PROCEDURES & MANEUVERS |
| 1. | | | | |
| 2. | | | | |
| 3. | | | | |
| 4. | | | | |
| 5. | | | | |
| 6. | | | | |
| 7. | | | | |
| 8. | | | | |
| 9. | | | | |
| 10. | | | | |
| TOTAL HOURS FOR SESSION | | | | |
| TOTAL FORWARD | | | | |
| TOTAL TO DATE | | | | |

## II. FLIGHT MAP

Month _____ Day _____ Year _____

**POSTFLIGHT NOTES / JOURNAL ENTRIES:**

I CERTIFY THAT THE FOREGOING ENTRIES ARE TRUE AND CORRECT:

PILOT: _____ SPOTTER: _____

# I. FLIGHT ID / NO.: _____

FLIGHT LOCATION: _____    Weather: _____

UNMANNED AIRCRAFT SYSTEM:                         CREW:

MANUFACTURER: _____            PILOT: _____

MODEL NUMBER: _____            SPOTTER: _____

PREFLIGHT CHECKLIST:

☐ Batteries Charged & Secure     ☐ Props OK & Tight            ☐ Compass Calibration
☐ Aircraft Hardware OK           ☐ Software / Firmware Update   ☐ Camera / FPV ON
☐ Equipment & Gear OK            ☐ Transmitter Control Power ON ☐ Satellite Connection
☐ Transmitter Controls OK        ☐ Aircraft Power ON            ☐ Applications / Other Systems ON

OPERATIONAL CONDITIONS & PREFLIGHT NOTES:

| SESSION FLIGHT INTERVALS | FLIGHT TIMES | | | REMARKS |
|---|---|---|---|---|
| | START | STOP | TOTAL | PROCEDURES & MANEUVERS |
| 1. | | | | |
| 2. | | | | |
| 3. | | | | |
| 4. | | | | |
| 5. | | | | |
| 6. | | | | |
| 7. | | | | |
| 8. | | | | |
| 9. | | | | |
| 10. | | | | |
| TOTAL HOURS FOR SESSION | | | | |
| TOTAL FORWARD | | | | |
| TOTAL TO DATE | | | | |

## II. FLIGHT MAP

Month _____ Day _____ Year _____

POSTFLIGHT NOTES / JOURNAL ENTRIES:

I CERTIFY THAT THE FOREGOING ENTRIES ARE TRUE AND CORRECT:

PILOT: _____ SPOTTER: _____

# I. FLIGHT ID / NO.: _____

FLIGHT LOCATION: _____     Weather: _____

UNMANNED AIRCRAFT SYSTEM:                     CREW:

MANUFACTURER: _____        PILOT: _____

MODEL NUMBER: _____        SPOTTER: _____

PREFLIGHT CHECKLIST:

☐ Batteries Charged & Secure   ☐ Props OK & Tight              ☐ Compass Calibration
☐ Aircraft Hardware OK         ☐ Software / Firmware Update     ☐ Camera / FPV ON
☐ Equipment & Gear OK          ☐ Transmitter Control Power ON   ☐ Satellite Connection
☐ Transmitter Controls OK      ☐ Aircraft Power ON              ☐ Applications / Other Systems ON

OPERATIONAL CONDITIONS & PREFLIGHT NOTES:

| SESSION FLIGHT INTERVALS | FLIGHT TIMES | | | REMARKS |
|---|---|---|---|---|
|  | START | STOP | TOTAL | PROCEDURES & MANEUVERS |
| 1. |  |  |  |  |
| 2. |  |  |  |  |
| 3. |  |  |  |  |
| 4. |  |  |  |  |
| 5. |  |  |  |  |
| 6. |  |  |  |  |
| 7. |  |  |  |  |
| 8. |  |  |  |  |
| 9. |  |  |  |  |
| 10. |  |  |  |  |
| TOTAL HOURS FOR SESSION |  |  |  |  |
| TOTAL FORWARD |  |  |  |  |
| TOTAL TO DATE |  |  |  |  |

## II. FLIGHT MAP

Month _____ Day _____ Year _____

**POSTFLIGHT NOTES / JOURNAL ENTRIES:**

I CERTIFY THAT THE FOREGOING ENTRIES ARE TRUE AND CORRECT:

PILOT: _____ SPOTTER: _____

# I. FLIGHT ID / NO.: _____

FLIGHT LOCATION: _____     Weather: _____

UNMANNED AIRCRAFT SYSTEM:                CREW:

MANUFACTURER: _____     PILOT: _____

MODEL NUMBER: _____     SPOTTER: _____

PREFLIGHT CHECKLIST:

☐ Batteries Charged & Secure    ☐ Props OK & Tight           ☐ Compass Calibration

☐ Aircraft Hardware OK          ☐ Software / Firmware Update  ☐ Camera / FPV ON

☐ Equipment & Gear OK           ☐ Transmitter Control Power ON  ☐ Satellite Connection

☐ Transmitter Controls OK       ☐ Aircraft Power ON          ☐ Applications / Other Systems ON

OPERATIONAL CONDITIONS & PREFLIGHT NOTES:

| SESSION FLIGHT INTERVALS | FLIGHT TIMES | | | REMARKS |
| --- | --- | --- | --- | --- |
| | START | STOP | TOTAL | PROCEDURES & MANEUVERS |
| 1. | | | | |
| 2. | | | | |
| 3. | | | | |
| 4. | | | | |
| 5. | | | | |
| 6. | | | | |
| 7. | | | | |
| 8. | | | | |
| 9. | | | | |
| 10. | | | | |
| TOTAL HOURS FOR SESSION | | | | |
| TOTAL FORWARD | | | | |
| TOTAL TO DATE | | | | |

## II. FLIGHT MAP

Month _____ Day _____ Year _____

POSTFLIGHT NOTES / JOURNAL ENTRIES:

I CERTIFY THAT THE FOREGOING ENTRIES ARE TRUE AND CORRECT:

PILOT: _____    SPOTTER: _____

# I. FLIGHT ID / NO.: _____

FLIGHT LOCATION: _____    Weather: _____

UNMANNED AIRCRAFT SYSTEM:                    CREW:

MANUFACTURER: _____       PILOT: _____

MODEL NUMBER: _____       SPOTTER: _____

PREFLIGHT CHECKLIST:

☐ Batteries Charged & Secure    ☐ Props OK & Tight              ☐ Compass Calibration
☐ Aircraft Hardware OK          ☐ Software / Firmware Update     ☐ Camera / FPV ON
☐ Equipment & Gear OK           ☐ Transmitter Control Power ON   ☐ Satellite Connection
☐ Transmitter Controls OK       ☐ Aircraft Power ON              ☐ Applications / Other Systems ON

OPERATIONAL CONDITIONS & PREFLIGHT NOTES:

| SESSION FLIGHT INTERVALS | FLIGHT TIMES | | | REMARKS |
| --- | --- | --- | --- | --- |
| | START | STOP | TOTAL | PROCEDURES & MANEUVERS |
| 1. | | | | |
| 2. | | | | |
| 3. | | | | |
| 4. | | | | |
| 5. | | | | |
| 6. | | | | |
| 7. | | | | |
| 8. | | | | |
| 9. | | | | |
| 10. | | | | |
| TOTAL HOURS FOR SESSION | | | | |
| TOTAL FORWARD | | | | |
| TOTAL TO DATE | | | | |

# II. FLIGHT MAP

Month _____ Day _____ Year _____

**POSTFLIGHT NOTES / JOURNAL ENTRIES:**

I CERTIFY THAT THE FOREGOING ENTRIES ARE TRUE AND CORRECT:

PILOT: _____    SPOTTER: _____

# I. FLIGHT ID / NO.: _____

FLIGHT LOCATION: _____    Weather: _____

## UNMANNED AIRCRAFT SYSTEM:                    CREW:

MANUFACTURER: _____          PILOT: _____

MODEL NUMBER: _____          SPOTTER: _____

## PREFLIGHT CHECKLIST:

| | | |
|---|---|---|
| ☐ Batteries Charged & Secure | ☐ Props OK & Tight | ☐ Compass Calibration |
| ☐ Aircraft Hardware OK | ☐ Software / Firmware Update | ☐ Camera / FPV ON |
| ☐ Equipment & Gear OK | ☐ Transmitter Control Power ON | ☐ Satellite Connection |
| ☐ Transmitter Controls OK | ☐ Aircraft Power ON | ☐ Applications / Other Systems ON |

## OPERATIONAL CONDITIONS & PREFLIGHT NOTES:

| SESSION FLIGHT INTERVALS | FLIGHT TIMES | | | REMARKS |
|---|---|---|---|---|
| | START | STOP | TOTAL | PROCEDURES & MANEUVERS |
| 1. | | | | |
| 2. | | | | |
| 3. | | | | |
| 4. | | | | |
| 5. | | | | |
| 6. | | | | |
| 7. | | | | |
| 8. | | | | |
| 9. | | | | |
| 10. | | | | |
| TOTAL HOURS FOR SESSION | | | | |
| TOTAL FORWARD | | | | |
| TOTAL TO DATE | | | | |

## II. FLIGHT MAP

Month _____ Day _____ Year _____

**POSTFLIGHT NOTES / JOURNAL ENTRIES:**

I CERTIFY THAT THE FOREGOING ENTRIES ARE TRUE AND CORRECT:

PILOT: _____ SPOTTER: _____

# I. FLIGHT ID / NO.: _____

FLIGHT LOCATION: _____   Weather: _____

UNMANNED AIRCRAFT SYSTEM:                    CREW:

MANUFACTURER: _____       PILOT: _____

MODEL NUMBER: _____       SPOTTER: _____

PREFLIGHT CHECKLIST:

☐ Batteries Charged & Secure    ☐ Props OK & Tight              ☐ Compass Calibration
☐ Aircraft Hardware OK          ☐ Software / Firmware Update     ☐ Camera / FPV ON
☐ Equipment & Gear OK           ☐ Transmitter Control Power ON   ☐ Satellite Connection
☐ Transmitter Controls OK       ☐ Aircraft Power ON              ☐ Applications / Other Systems ON

OPERATIONAL CONDITIONS & PREFLIGHT NOTES:

| SESSION FLIGHT INTERVALS | FLIGHT TIMES | | | REMARKS |
| --- | --- | --- | --- | --- |
| | START | STOP | TOTAL | PROCEDURES & MANEUVERS |
| 1. | | | | |
| 2. | | | | |
| 3. | | | | |
| 4. | | | | |
| 5. | | | | |
| 6. | | | | |
| 7. | | | | |
| 8. | | | | |
| 9. | | | | |
| 10. | | | | |
| TOTAL HOURS FOR SESSION | | | | |
| TOTAL FORWARD | | | | |
| TOTAL TO DATE | | | | |

## II. FLIGHT MAP

Month _____ Day _____ Year _____

**POSTFLIGHT NOTES / JOURNAL ENTRIES:**

I CERTIFY THAT THE FOREGOING ENTRIES ARE TRUE AND CORRECT:

PILOT: _____   SPOTTER: _____

# I. FLIGHT ID / NO.: _____

FLIGHT LOCATION: _____     Weather: _____

UNMANNED AIRCRAFT SYSTEM:                  CREW:

MANUFACTURER: _____        PILOT: _____

MODEL NUMBER: _____        SPOTTER: _____

PREFLIGHT CHECKLIST:

☐ Batteries Charged & Secure    ☐ Props OK & Tight            ☐ Compass Calibration
☐ Aircraft Hardware OK          ☐ Software / Firmware Update  ☐ Camera / FPV ON
☐ Equipment & Gear OK           ☐ Transmitter Control Power ON ☐ Satellite Connection
☐ Transmitter Controls OK       ☐ Aircraft Power ON           ☐ Applications / Other Systems ON

OPERATIONAL CONDITIONS & PREFLIGHT NOTES:

| SESSION FLIGHT INTERVALS | FLIGHT TIMES | | | REMARKS |
| --- | --- | --- | --- | --- |
| | START | STOP | TOTAL | PROCEDURES & MANEUVERS |
| 1. | | | | |
| 2. | | | | |
| 3. | | | | |
| 4. | | | | |
| 5. | | | | |
| 6. | | | | |
| 7. | | | | |
| 8. | | | | |
| 9. | | | | |
| 10. | | | | |
| TOTAL HOURS FOR SESSION | | | | |
| TOTAL FORWARD | | | | |
| TOTAL TO DATE | | | | |

## II. FLIGHT MAP

Month _____ Day _____ Year _____

**POSTFLIGHT NOTES / JOURNAL ENTRIES:**

I CERTIFY THAT THE FOREGOING ENTRIES ARE TRUE AND CORRECT:

PILOT: _____     SPOTTER: _____

# I. FLIGHT ID / NO.: _____

FLIGHT LOCATION: _____     Weather: _____

## UNMANNED AIRCRAFT SYSTEM:                      CREW:

MANUFACTURER: _____          PILOT: _____

MODEL NUMBER: _____          SPOTTER: _____

## PREFLIGHT CHECKLIST:

☐ Batteries Charged & Secure   ☐ Props OK & Tight              ☐ Compass Calibration
☐ Aircraft Hardware OK         ☐ Software / Firmware Update    ☐ Camera / FPV ON
☐ Equipment & Gear OK          ☐ Transmitter Control Power ON  ☐ Satellite Connection
☐ Transmitter Controls OK      ☐ Aircraft Power ON             ☐ Applications / Other Systems ON

## OPERATIONAL CONDITIONS & PREFLIGHT NOTES:

| SESSION FLIGHT INTERVALS | FLIGHT TIMES | | | REMARKS |
| --- | --- | --- | --- | --- |
| | START | STOP | TOTAL | PROCEDURES & MANEUVERS |
| 1. | | | | |
| 2. | | | | |
| 3. | | | | |
| 4. | | | | |
| 5. | | | | |
| 6. | | | | |
| 7. | | | | |
| 8. | | | | |
| 9. | | | | |
| 10. | | | | |
| TOTAL HOURS FOR SESSION | | | | |
| TOTAL FORWARD | | | | |
| TOTAL TO DATE | | | | |

## II. FLIGHT MAP

Month _____ Day _____ Year _____

POSTFLIGHT NOTES / JOURNAL ENTRIES:

I CERTIFY THAT THE FOREGOING ENTRIES ARE TRUE AND CORRECT:

PILOT: _____     SPOTTER: _____

# I. FLIGHT ID / NO.: _____

FLIGHT LOCATION: _____          Weather: _____

UNMANNED AIRCRAFT SYSTEM:                         CREW:

MANUFACTURER: _____          PILOT: _____

MODEL NUMBER: _____          SPOTTER: _____

PREFLIGHT CHECKLIST:

☐ Batteries Charged & Secure    ☐ Props OK & Tight               ☐ Compass Calibration
☐ Aircraft Hardware OK          ☐ Software / Firmware Update     ☐ Camera / FPV ON
☐ Equipment & Gear OK           ☐ Transmitter Control Power ON   ☐ Satellite Connection
☐ Transmitter Controls OK       ☐ Aircraft Power ON              ☐ Applications / Other Systems ON

OPERATIONAL CONDITIONS & PREFLIGHT NOTES:

| SESSION FLIGHT INTERVALS | FLIGHT TIMES | | | REMARKS |
| --- | --- | --- | --- | --- |
| | START | STOP | TOTAL | PROCEDURES & MANEUVERS |
| 1. | | | | |
| 2. | | | | |
| 3. | | | | |
| 4. | | | | |
| 5. | | | | |
| 6. | | | | |
| 7. | | | | |
| 8. | | | | |
| 9. | | | | |
| 10. | | | | |
| TOTAL HOURS FOR SESSION | | | | |
| TOTAL FORWARD | | | | |
| TOTAL TO DATE | | | | |

## II. FLIGHT MAP

Month _____ Day _____ Year _____

(grid of dots)

**POSTFLIGHT NOTES / JOURNAL ENTRIES:**

I CERTIFY THAT THE FOREGOING ENTRIES ARE TRUE AND CORRECT:

PILOT: _____     SPOTTER: _____

# I. FLIGHT ID / NO.: _____

FLIGHT LOCATION: _____     Weather: _____

UNMANNED AIRCRAFT SYSTEM:                    CREW:

MANUFACTURER: _____        PILOT: _____

MODEL NUMBER: _____        SPOTTER: _____

PREFLIGHT CHECKLIST:

☐ Batteries Charged & Secure    ☐ Props OK & Tight              ☐ Compass Calibration
☐ Aircraft Hardware OK          ☐ Software / Firmware Update     ☐ Camera / FPV ON
☐ Equipment & Gear OK           ☐ Transmitter Control Power ON   ☐ Satellite Connection
☐ Transmitter Controls OK       ☐ Aircraft Power ON              ☐ Applications / Other Systems ON

OPERATIONAL CONDITIONS & PREFLIGHT NOTES:

| SESSION FLIGHT INTERVALS | FLIGHT TIMES | | | REMARKS |
| --- | --- | --- | --- | --- |
| | START | STOP | TOTAL | PROCEDURES & MANEUVERS |
| 1. | | | | |
| 2. | | | | |
| 3. | | | | |
| 4. | | | | |
| 5. | | | | |
| 6. | | | | |
| 7. | | | | |
| 8. | | | | |
| 9. | | | | |
| 10. | | | | |
| TOTAL HOURS FOR SESSION | | | | |
| TOTAL FORWARD | | | | |
| TOTAL TO DATE | | | | |

## II. FLIGHT MAP

Month _____ Day _____ Year _____

POSTFLIGHT NOTES / JOURNAL ENTRIES:

I CERTIFY THAT THE FOREGOING ENTRIES ARE TRUE AND CORRECT:

PILOT: _____ SPOTTER: _____

# I. FLIGHT ID / NO.: _____

FLIGHT LOCATION: _____     Weather: _____

UNMANNED AIRCRAFT SYSTEM:                       CREW:

MANUFACTURER: _____         PILOT: _____

MODEL NUMBER: _____         SPOTTER: _____

## PREFLIGHT CHECKLIST:

☐ Batteries Charged & Secure     ☐ Props OK & Tight              ☐ Compass Calibration

☐ Aircraft Hardware OK           ☐ Software / Firmware Update    ☐ Camera / FPV ON

☐ Equipment & Gear OK            ☐ Transmitter Control Power ON  ☐ Satellite Connection

☐ Transmitter Controls OK        ☐ Aircraft Power ON             ☐ Applications / Other Systems ON

## OPERATIONAL CONDITIONS & PREFLIGHT NOTES:

| SESSION FLIGHT INTERVALS | FLIGHT TIMES | | | REMARKS |
|---|---|---|---|---|
| | START | STOP | TOTAL | PROCEDURES & MANEUVERS |
| 1. | | | | |
| 2. | | | | |
| 3. | | | | |
| 4. | | | | |
| 5. | | | | |
| 6. | | | | |
| 7. | | | | |
| 8. | | | | |
| 9. | | | | |
| 10. | | | | |
| TOTAL HOURS FOR SESSION | | | | |
| TOTAL FORWARD | | | | |
| TOTAL TO DATE | | | | |

## II. FLIGHT MAP

Month _____ Day _____ Year _____

**POSTFLIGHT NOTES / JOURNAL ENTRIES:**

I CERTIFY THAT THE FOREGOING ENTRIES ARE TRUE AND CORRECT:

PILOT: _____     SPOTTER: _____

# I. FLIGHT ID / NO.: _____

FLIGHT LOCATION: _____    Weather: _____

UNMANNED AIRCRAFT SYSTEM:                    CREW:

MANUFACTURER: _____   PILOT: _____

MODEL NUMBER: _____   SPOTTER: _____

PREFLIGHT CHECKLIST:

☐ Batteries Charged & Secure      ☐ Props OK & Tight            ☐ Compass Calibration
☐ Aircraft Hardware OK            ☐ Software / Firmware Update   ☐ Camera / FPV ON
☐ Equipment & Gear OK             ☐ Transmitter Control Power ON ☐ Satellite Connection
☐ Transmitter Controls OK         ☐ Aircraft Power ON           ☐ Applications / Other Systems ON

OPERATIONAL CONDITIONS & PREFLIGHT NOTES:

| SESSION FLIGHT INTERVALS | FLIGHT TIMES | | | REMARKS |
| --- | --- | --- | --- | --- |
| | START | STOP | TOTAL | PROCEDURES & MANEUVERS |
| 1. | | | | |
| 2. | | | | |
| 3. | | | | |
| 4. | | | | |
| 5. | | | | |
| 6. | | | | |
| 7. | | | | |
| 8. | | | | |
| 9. | | | | |
| 10. | | | | |
| TOTAL HOURS FOR SESSION | | | | |
| TOTAL FORWARD | | | | |
| TOTAL TO DATE | | | | |

## II. FLIGHT MAP

Month _____ Day _____ Year _____

POSTFLIGHT NOTES / JOURNAL ENTRIES:

I CERTIFY THAT THE FOREGOING ENTRIES ARE TRUE AND CORRECT:

PILOT: _____     SPOTTER: _____

# I. FLIGHT ID / NO.: _____

FLIGHT LOCATION: _____     Weather: _____

UNMANNED AIRCRAFT SYSTEM:                  CREW:

MANUFACTURER: _____        PILOT: _____

MODEL NUMBER: _____        SPOTTER: _____

PREFLIGHT CHECKLIST:

☐ Batteries Charged & Secure    ☐ Props OK & Tight            ☐ Compass Calibration
☐ Aircraft Hardware OK          ☐ Software / Firmware Update  ☐ Camera / FPV ON
☐ Equipment & Gear OK           ☐ Transmitter Control Power ON ☐ Satellite Connection
☐ Transmitter Controls OK       ☐ Aircraft Power ON           ☐ Applications / Other Systems ON

OPERATIONAL CONDITIONS & PREFLIGHT NOTES:

| SESSION FLIGHT INTERVALS | FLIGHT TIMES | | | REMARKS |
| --- | --- | --- | --- | --- |
| | START | STOP | TOTAL | PROCEDURES & MANEUVERS |
| 1. | | | | |
| 2. | | | | |
| 3. | | | | |
| 4. | | | | |
| 5. | | | | |
| 6. | | | | |
| 7. | | | | |
| 8. | | | | |
| 9. | | | | |
| 10. | | | | |
| TOTAL HOURS FOR SESSION | | | | |
| TOTAL FORWARD | | | | |
| TOTAL TO DATE | | | | |

# II. FLIGHT MAP

Month _____ Day _____ Year _____

**POSTFLIGHT NOTES / JOURNAL ENTRIES:**

I CERTIFY THAT THE FOREGOING ENTRIES ARE TRUE AND CORRECT:

PILOT: _____ SPOTTER: _____

# I. FLIGHT ID / NO.: _____

FLIGHT LOCATION: _____     Weather: _____

UNMANNED AIRCRAFT SYSTEM:                    CREW:

MANUFACTURER: _____          PILOT: _____

MODEL NUMBER: _____          SPOTTER: _____

PREFLIGHT CHECKLIST:

☐ Batteries Charged & Secure    ☐ Props OK & Tight              ☐ Compass Calibration
☐ Aircraft Hardware OK          ☐ Software / Firmware Update     ☐ Camera / FPV ON
☐ Equipment & Gear OK           ☐ Transmitter Control Power ON   ☐ Satellite Connection
☐ Transmitter Controls OK       ☐ Aircraft Power ON              ☐ Applications / Other Systems ON

OPERATIONAL CONDITIONS & PREFLIGHT NOTES:

| SESSION FLIGHT INTERVALS | FLIGHT TIMES | | | REMARKS |
| --- | --- | --- | --- | --- |
| | START | STOP | TOTAL | PROCEDURES & MANEUVERS |
| 1. | | | | |
| 2. | | | | |
| 3. | | | | |
| 4. | | | | |
| 5. | | | | |
| 6. | | | | |
| 7. | | | | |
| 8. | | | | |
| 9. | | | | |
| 10. | | | | |
| TOTAL HOURS FOR SESSION | | | | |
| TOTAL FORWARD | | | | |
| TOTAL TO DATE | | | | |

## II. FLIGHT MAP

Month _____ Day _____ Year _____

**POSTFLIGHT NOTES / JOURNAL ENTRIES:**

I CERTIFY THAT THE FOREGOING ENTRIES ARE TRUE AND CORRECT:

PILOT: _____     SPOTTER: _____

# I. FLIGHT ID / NO.: _____

FLIGHT LOCATION: _____  Weather: _____

## UNMANNED AIRCRAFT SYSTEM:                    CREW:

MANUFACTURER: _____     PILOT: _____

MODEL NUMBER: _____     SPOTTER: _____

## PREFLIGHT CHECKLIST:

☐ Batteries Charged & Secure    ☐ Props OK & Tight              ☐ Compass Calibration
☐ Aircraft Hardware OK          ☐ Software / Firmware Update    ☐ Camera / FPV ON
☐ Equipment & Gear OK           ☐ Transmitter Control Power ON  ☐ Satellite Connection
☐ Transmitter Controls OK       ☐ Aircraft Power ON             ☐ Applications / Other Systems ON

## OPERATIONAL CONDITIONS & PREFLIGHT NOTES:

| SESSION FLIGHT INTERVALS | FLIGHT TIMES | | | REMARKS |
|---|---|---|---|---|
| | START | STOP | TOTAL | PROCEDURES & MANEUVERS |
| 1. | | | | |
| 2. | | | | |
| 3. | | | | |
| 4. | | | | |
| 5. | | | | |
| 6. | | | | |
| 7. | | | | |
| 8. | | | | |
| 9. | | | | |
| 10. | | | | |
| TOTAL HOURS FOR SESSION | | | | |
| TOTAL FORWARD | | | | |
| TOTAL TO DATE | | | | |

## II. FLIGHT MAP

Month _____ Day _____ Year _____

**POSTFLIGHT NOTES / JOURNAL ENTRIES:**

I CERTIFY THAT THE FOREGOING ENTRIES ARE TRUE AND CORRECT:

PILOT: _____     SPOTTER: _____

# I. FLIGHT ID / NO.: _____

FLIGHT LOCATION: _____     Weather: _____

UNMANNED AIRCRAFT SYSTEM:                     CREW:

MANUFACTURER: _____         PILOT: _____

MODEL NUMBER: _____         SPOTTER: _____

PREFLIGHT CHECKLIST:

☐ Batteries Charged & Secure      ☐ Props OK & Tight              ☐ Compass Calibration
☐ Aircraft Hardware OK            ☐ Software / Firmware Update     ☐ Camera / FPV ON
☐ Equipment & Gear OK             ☐ Transmitter Control Power ON   ☐ Satellite Connection
☐ Transmitter Controls OK         ☐ Aircraft Power ON              ☐ Applications / Other Systems ON

OPERATIONAL CONDITIONS & PREFLIGHT NOTES:

| SESSION FLIGHT INTERVALS | FLIGHT TIMES | | | REMARKS |
| --- | --- | --- | --- | --- |
| | START | STOP | TOTAL | PROCEDURES & MANEUVERS |
| 1. | | | | |
| 2. | | | | |
| 3. | | | | |
| 4. | | | | |
| 5. | | | | |
| 6. | | | | |
| 7. | | | | |
| 8. | | | | |
| 9. | | | | |
| 10. | | | | |
| TOTAL HOURS FOR SESSION | | | | |
| TOTAL FORWARD | | | | |
| TOTAL TO DATE | | | | |

## II. FLIGHT MAP

Month _____ Day _____ Year _____

**POSTFLIGHT NOTES / JOURNAL ENTRIES:**

I CERTIFY THAT THE FOREGOING ENTRIES ARE TRUE AND CORRECT:

PILOT: _____ SPOTTER: _____

# I. FLIGHT ID / NO.: _____

FLIGHT LOCATION: _____    Weather: _____

UNMANNED AIRCRAFT SYSTEM:                        CREW:

MANUFACTURER: _____        PILOT: _____

MODEL NUMBER: _____         SPOTTER: _____

PREFLIGHT CHECKLIST:

☐ Batteries Charged & Secure    ☐ Props OK & Tight              ☐ Compass Calibration
☐ Aircraft Hardware OK          ☐ Software / Firmware Update     ☐ Camera / FPV ON
☐ Equipment & Gear OK           ☐ Transmitter Control Power ON   ☐ Satellite Connection
☐ Transmitter Controls OK       ☐ Aircraft Power ON              ☐ Applications / Other Systems ON

OPERATIONAL CONDITIONS & PREFLIGHT NOTES:

| SESSION FLIGHT INTERVALS | FLIGHT TIMES | | | REMARKS |
|---|---|---|---|---|
| | START | STOP | TOTAL | PROCEDURES & MANEUVERS |
| 1. | | | | |
| 2. | | | | |
| 3. | | | | |
| 4. | | | | |
| 5. | | | | |
| 6. | | | | |
| 7. | | | | |
| 8. | | | | |
| 9. | | | | |
| 10. | | | | |
| TOTAL HOURS FOR SESSION | | | | |
| TOTAL FORWARD | | | | |
| TOTAL TO DATE | | | | |

## II. FLIGHT MAP

Month _____ Day _____ Year _____

POSTFLIGHT NOTES / JOURNAL ENTRIES:

I CERTIFY THAT THE FOREGOING ENTRIES ARE TRUE AND CORRECT:

PILOT: _____     SPOTTER: _____

# I. FLIGHT ID / NO.: _____

FLIGHT LOCATION: _____     Weather: _____

UNMANNED AIRCRAFT SYSTEM:                    CREW:

MANUFACTURER: _____     PILOT: _____

MODEL NUMBER: _____     SPOTTER: _____

## PREFLIGHT CHECKLIST:

☐ Batteries Charged & Secure     ☐ Props OK & Tight     ☐ Compass Calibration

☐ Aircraft Hardware OK     ☐ Software / Firmware Update     ☐ Camera / FPV ON

☐ Equipment & Gear OK     ☐ Transmitter Control Power ON     ☐ Satellite Connection

☐ Transmitter Controls OK     ☐ Aircraft Power ON     ☐ Applications / Other Systems ON

## OPERATIONAL CONDITIONS & PREFLIGHT NOTES:

| SESSION FLIGHT INTERVALS | FLIGHT TIMES | | | REMARKS |
| --- | --- | --- | --- | --- |
| | START | STOP | TOTAL | PROCEDURES & MANEUVERS |
| 1. | | | | |
| 2. | | | | |
| 3. | | | | |
| 4. | | | | |
| 5. | | | | |
| 6. | | | | |
| 7. | | | | |
| 8. | | | | |
| 9. | | | | |
| 10. | | | | |
| TOTAL HOURS FOR SESSION | | | | |
| TOTAL FORWARD | | | | |
| TOTAL TO DATE | | | | |

## II. FLIGHT MAP

Month _____ Day _____ Year _____

**POSTFLIGHT NOTES / JOURNAL ENTRIES:**

I CERTIFY THAT THE FOREGOING ENTRIES ARE TRUE AND CORRECT:

PILOT: _____ SPOTTER: _____

Made in the USA
Coppell, TX
10 November 2022

86101022R00063